Lab Safety & Equipment

TEACHING CHEMISTRY IN A DIVERSIFIED CLASSROOM

BOOK 1

By Sammie Jacobs

Copyright

All rights reserved under International and Pan-American Copyright Conventions. By payment of the required fees, you have been granted the non-exclusive, non-transferable right to use the materials provided for the preparation of lessons and during the direct instruction of students.

LAB SAFETY & EQUIPMENT. Copyright 2020 ©

Student worksheets and assessments may be duplicated for classroom use, the number not to exceed the number of students in each class. Notice of copyright must appear on all copies as provided on the page.

Activity kits duplicated and assembled as directed should display the copyright markings as prescribed in the directions and on the materials.

No other part of this text may be reproduced, up-loaded, displayed, shared, transmitted, downloaded, decompiled, reverse engineered, or stored in or introduced into any information storage and retrieval system, in any form or by any means, whether electronic or mechanical, now known or hereinafter invented, without the express written permission of Lavish Publishing, LLC.

First Edition
2020 Lavish Publishing, LLC
Teaching Chemistry in a Diversified Classroom book 1
All Rights Reserved
Published in the United States by Lavish Publishing, LLC, Midland, Texas
Select Illustrations by Sanghamitra Dasgupta
Cover Design by: Victor R. Sosa
Cover Images: CanStock
Paperback edition
ISBN: 978-1-64900-000-2
www.LavishPublishing.com

FOREWORD

Teachers, parents, and educators of all kinds – welcome to my classroom!

Here within these pages, and in fact this entire series, are lessons and activities that I have created and used with my own students over the years presented in an easy-to-use format. Through trial and error, hardship and success, I have learned how to present the difficult world of introductory chemistry to students in scalable terms, ways that lend themselves to a wide variety of learners.

So, if you teach in a small class, large class, average class, have ESL or SPED, or even if you are homeschooling, you will find the flexibility of my lesson plans tailor made for you. Each unit includes a complete calendar plan, lessons for each classroom day for forty-five minutes of instruction each, and even formative and summative assessments to check for learning. Simply use the lessons, scale them as need be, and augment whenever you like.

Be sure to join my Teaching Chemistry PLC on Facebook for even more ideas and sharing, and of course, I hope for the best with you and your students!

Sammie Jacobs

TABLE OF CONTENTS

Lab Safety & Equipment Unit Tools . Pages 1 – 5
Calendar Key and Usage
Lab Safety and Equipment Guide Calendar
Lab Safety and Equipment Blank Calendar
Lab Safety and Equipment Vocabulary

Lesson Plans . . . . .Pages 6 – 34
Meet and Greet
Tool Set Up
Reading Strategy & Safety Data Sheets
Lab Safety Project Building
Lab Safety Project Presentations
Lab Equipment
Scientific Method
Video Day
Lab Safety and Equipment Review

Assessments . . . . .Pages 35 – 44
Equipment Quiz
Scientific Method Quiz
Safety Hazard Quiz
Lab Safety Vocabulary Quizzes
Lab Safety Unit Exam
Answer Keys

Unit Calendar

The **Unit Calendar** is an important organizational tool.

It features a list of vocabulary terms (top section) for the unit, the plan of days (center section), and the student learning goals (bottom section).

A **completed teacher copy** of the calendar with the unit overview is provided. On this version, days are not written as dates, but as lesson days. However, on my copy that I use in my classroom, these have been laid to the school or district calendar and those day numbers are replaced with actual meeting calendar dates. There are five in a row, which is a work week, and days that we do not meet or have instruction time are crossed off.

You will notice that the unit calendars are always a whole number of weeks long and every day has a lesson planned, but often you will have holidays or non-class days. When you have less days, combine lessons where appropriate, or omit lessons that are not included on the student goals and exam if necessary. We also only meet for forty-five minutes per day, so the lessons are designed to fit within that time frame. If you have longer class periods, you may decide to augment the days with additional activities, such as vocabulary games.

The **blank student copy** can be used to customize the calendar to match your available days. It can also be copied and distributed to students, along with the vocabulary lists with definitions.

Calendar ideas:

1. Have students fill in the calendar daily with daily topic as they begin class as a way of making them aware of what the day's lesson will entail.
2. Have students fill out the calendar at the start of a new unit so they have the plan ahead of time and can be proactive with their learning.
3. Give students a completed copy with the appropriate dates and topics at the start of a new unit to save time.
4. Have students highlight each day and the corresponding terms on the vocab list with a different color (color coordinating them) for reviewing later.
5. Have students use the calendar to record assignments for each day and highlight them as they are completed and submitted.
6. Have the students write a question for each day to reflect upon later.

Unit	Lab Safety & Equipment	Month

Vocabulary

accident	eyewash fountain	observation
beaker	fire blanket	pipette
conservation	fire extinguisher	proactive
constant	flammability	problem question
control	Florence flask	radioactivity
corrosiveness	graduated cylinder	reagent bottle
dependent variable	hazard	recycle
disposal	hypothesis	repeated trials
emergency button	independent variable	risk management
empirical evidence	lab apron	safety goggles
Erlenmeyer flask	lab journal	safety shower
experimental	logical reasoning	variable

Daily Agenda

M 1	T 2	W 3	Th 4	F 5
Meet & Greet	Journal and Google Classroom Set Up Day	Reading Strategy Modeling Safety Data Sheet	Safety Project	Safety Project Presentations

M 6	T 7	W 8	Th 9	F 10
Lab Equipment	Scientific Method Lab	Video Day	Flex Day and Unit Review	Safety & Equipment Exam

Learning Goals

Target Concept	none	weak	solid
Lab Safety Rules - identify rules and examples / non-examples			
Lab Equipment - be able to identify the basic equipment			
Hazards - what are they, precautions, emergencies & disposal			
Scientific Method - dependent, independent, hypothesis, conclusions			

© Lavish Publishing, LLC

Unit **Lab Safety & Equipment** Month _____

Vocabulary

accident	eyewash fountain	observation
beaker	fire blanket	pipette
conservation	fire extinguisher	proactive
constant	flammability	problem question
control	Florence flask	radioactivity
corrosiveness	graduated cylinder	reagent bottle
dependent variable	hazard	recycle
disposal	hypothesis	repeated trials
emergency button	independent variable	risk management
empirical evidence	lab apron	safety goggles
Erlenmeyer flask	lab journal	safety shower
experimental	logical reasoning	variable

Daily Agenda

M	T	W	Th	F

M	T	W	Th	F

Learning Goals

Target Concept	none	weak	solid
Lab Safety Rules - identify rules and examples / non-examples			
Lab Equipment - be able to identify the basic equipment			
Hazards - what are they, precautions, emergencies & disposal			
Scientific Method - dependent, independent, hypothesis, conclusions			

© Lavish Publishing, LLC

Lab Safety and Equipment Vocabulary

word	definition
accident	An ____ is unexpected and causes property damage or personal injury.
beaker	A common piece of equipment is the ___, which is often made of glass, used to heat liquids, and for holding chemicals during lab procedures.
conservation	____ is to use our resources sparingly or in a non-wasteful manner. It also refers to some of our laws in science that state that certain things cannot be created or destroyed, such as matter or energy.
constant	The things we want to hold the same in an experiment are called _____.
control	The ____ in the experiment is set aside for comparison and not altered.
corrosiveness	A chemical property, ____ can cause skin irritation and burns, as well as making stains on and holes in clothing and other items.
dependent	The ____ variable reacts in the experiment and is observed; the outcome.
disposal	___ means to throw something away or getting rid of it - we always must do this in the proper manner to prevent contaminating our water supply and landfills.
emergency button	The ____ should only be pushed in an emergency, as it will shut off all water, gas, and electricity to lab tables, and activate the vent system.
empirical evidence	___ is evidence that is found by a direct observation, experience, or situation, which makes it more believable.
Erlenmeyer flask	An ____ has a short neck from the mouth to a body that flares out from a narrow top to a wide base. Used for holding and mixing chemicals.
experimental	An _____ value is found in a lab.
eyewash fountain	If chemicals splash in someone's face, they should use the ___ for 15 minutes to ensure the chemical is flushed out completely.
fire blanket	The best method for helping a person on fire is to throw the ____ over them and smother it.
fire extinguisher	In case of a fire, we could use the ____ to put it out by pulling the pin and spraying the fire in a sweeping motion.
flammability	A chemical property, ____ is the ability to burn or catch fire.
Florence flask	A ___ is a flask with a narrow tube from the mouth down to a round base with a flat bottom for sitting on a table.

graduated cylinder	A ___ is a piece of laboratory glassware or plastic ware used to accurately measure out volumes of chemicals or other liquids. They are more accurate or precise for this purpose than flasks or beakers.
hazard	A ___ is something that could be dangerous and should be handled with caution.
hypothesis	___ is an educated guess.
independent	The ___ variable is manipulated and tested in an experiment.
lab apron	The piece of lab equipment that protects us and our clothing is a ___.
lab journal	A ___ is used to record data and observations for lab activities.
logical reasoning	___ would be looking at all of the facts and lining up those facts in such a way that you come to a documented conclusion.
observation	An ___ is something that we notice or make note of while performing a task or experiment.
pipette	A ___ is a laboratory instrument used to transport a measured volume of liquid, similar to an eye-dropper. They are very accurate or precise.
proactive	To be ___ is to prepare for something ahead of time.
problem question	After making an observation at the beginning of the scientific process, we ask a ___.
radioactivity	Materials that pose a ___ hazard are marked with the symbol to signify that there may be harmful particles emitted from them. You may also see this symbol in places where X-rays are made, such as a hospital.
reagent bottle	A ___ is used to hold chemicals in storage - one should NEVER pour things into or stick items inside of them to prevent contamination.
recycle	When we ___, we can help save money and / or the environment by reducing the amount of waste in our landfills and the use of our natural resources used to produce new products.
repeated trials	___ are done to obtain more accurate data.
risk management	___ lowers accidents through team efforts to be safe.
safety goggles	___ are worn to protect our eyes when we are using chemicals, glassware, or fire (all labs in chemistry probably require them).
safety shower	A piece of lab safety equipment that can dump water on your entire body in case of an emergency is the ___.
variable	A ___ is anything that can be changed or changes in an experiment.

© Lavish Publishing, LLC

Key to Lesson Plans

Topic: coordinated to the unit calendar - what we are learning about today?
Day: coordinated to the calendar - sequence of presentation out of total available
Unit: coordinated to the unit calendar - unit the lesson falls under

Learning Target:
A postable **learning objective** that gives an outline of what the day's lesson will cover and **artifact** the students will produce for evaluation. These can be used as is or modified to suit your student output or to focus on a different aspect of the lesson, as these often only cover one part of a layered lesson and group of activities.

Student Goals:
At the bottom of the calendar, the students have a list of **goals** they want to meet by the end of the unit, which are covered on the **unit exam**. This section coordinates which goals are being addressed during this lesson. This section also lists the **vocabulary terms** from the calendar that will be defined or needed during the lesson so they can be connected, pre-taught, or directed to study afterwards.

Agenda:
A postable list of what activities this day will include. They are divided into two main types of instruction:

Lecture **(L)**, where the teacher is providing direct instruction and the students are actively listening, taking notes and providing feedback at given intervals.

Activities **(A)**, where the student is producing work of some kind and the teacher is observing and providing support when needed.

Student Materials:
A list of materials that each student, pair, or group, will need to complete the lesson. They will need to be prepared ahead of time and be ready to **pick up** as students enter the room or to be **distributed** at the appropriate time during the lesson.

Generally, if it is a **one-to-one item**, I have them ready and students pick them up as they come in to save time. **Paired and Group materials** can be handed out or placed in a location to have a student go and get while the teacher is preparing some other part or completing a task during transition.

Props:
These are items you generally only need one of and can be hidden to pull out at the appropriate time or placed on a table that the students can visit before and after the lesson.

They are support items that deepen the understanding of the lecture and either generate questions or provide answers.

Having props is vital for a wide range of learners, so do not feel limited to what I have and use – explore and add anything that you want to aid in this process, as all students benefit. Save your props as you build them, as they often are used in future lessons to bring concepts forward and tie ideas and understanding together.

I have a set of props that stay on my front table throughout the year, as I pick them up and refer to them often: a plain bottle of water that is labeled H_2O, a bottle that is half cooking oil and half water, a bottle that is water with about a gram of dirt in it, and a bottle that is water with about a gram of corn starch in it. Other props are added and removed as needed.

Actions and Rationale:
These are the key points of the lesson and the reasoning behind what is being said or done. All activities will have a separate directions page if needed, as well as a reproduceable student copy and directions for building all props and student materials.

During each unit, we use a variety of learning and presentation styles that include having the students listen, speak, read and write. There are also a variety of study and practice skills woven into the lessons that students can learn to use in and outside of their Chemistry class.

Topic: Meet & Greet
Day: 1
Unit: Lab Safety & Equipment

Learning Target:
I can learn about the expectations in Chemistry class and write personal goals for myself.

Student Goals:
Students should feel ownership of their outcomes for the class – writing a goal of what they want to accomplish will give them something to work towards. I let them write their own, as it should be personal to them.

Agenda:
A – Name Game
L – Class Expectations
A – My Goals for the Year

Student Materials:
3 x 5 notecard or slip of paper

Props:
Write the instructions for the goals on the board or paper / power point to display on the board.

Actions and Rationale:
Every activity during the first unit is an opportunity to teach your students how your classroom works. I have a table set up with what the students will pick up each day as they enter. On this first day, I put the cards on the table and as I greet them at the door, I have them pick up a card (or slip of paper if you choose).

During class, I follow the **agenda**, starting with a few minutes of our first **activity - learning student names**. If the class is too large, you can save that for the end to get as many as possible in the time allowed. There are a variety of games that will work for this, but generally, I do one of these:

"Name Rotation" is where the first person says their name. The second person says the first, then their own name. The third names person one, person two, then themselves. And on it goes until the last person names everyone in the room in order.

"Tell me more" is one I like to do with a smaller group, where each student has a minute or two to speak. I ask them a few personal things about them to get to know them better, such as "What do you like to do after school?" I make notes on a seating chart with their name and something about them as we go, such as if they have a job, are in the band, on the football team, etc.

"I want to be" is a chance to get to know what your students are thinking about their future, which also gives you something to help remember them by, same as the tell me more version. Choose one of these or make your own – the possibilities are endless.

For the **Lecture**, I go over any pressing rules that the students will need to know or follow right away, such as how I handle tardies or absences.

For the final **activity**, the students will end class (or you can start class as a warm up) by writing a short goal for themselves – this can be a class goal, a life goal such as career thoughts, or a combination of the two. Or if you start class with it, they can write about "What you do after school," or "I want to be" and be ready for the activity.

Topic: Set Up Day
Day: 2
Unit: Lab Safety

Learning Target:
I can learn to be organized and use my resources for Chemistry class by setting up my journal and joining Google Classroom.

Student Goals:
Student's in my class must learn early on that required work is non-negotiable. I provide multiple options or tools, but the outcomes must be achieved for students to be successful this year, which includes learning to follow directions and be self-reliant. Today, they begin building their tools and formulating how they will access them.

Agenda:
A – Pick up a Journal and put your name on the front
L/A – Joining Google Classroom and Set up Journal

Student Materials:
Journal for each student
Sharpies and glue to share
Copies of Class Rules
Copies of Safety Contract
Copies of Safety Unit Calendar
Copies of Safety Unit Vocab
Copies of Periodic Table
6 x 9 envelope for each student

Props:
Google Glass access code to display
Teacher Journal to demo
Crates or Cabinets for Journal Storage

Actions and Rationale:
Virtual Classroom - Set up your virtual classroom beforehand – I make a folder for each unit and preload as much as I can to save myself time later in the year. Typically, the calendar, vocab, and my power points and notes are all in there before class ever begins. If you are not able to work that far ahead, you can always load as you go, which is how I handle the rest of my materials. Finally, take pictures of things you put on the board and upload

those as well – it helps with students who are absent and working to catch up, and it makes great review material before exams.

I use google classroom, but there are others if you would prefer something else. I also purchase journals for all of my students at a superstore for twenty-five cents each – they are generally inexpensive and well worth the investment to ensure the students all have them on time. I have also found it helps to have them all the same style and type, which buying them myself ensures will happen. We use 70 page wide rule, and there will often be pictures of the journal notes to go with every unit if you care to use them – or tweak them as you please.

As for accessing our virtual classroom, at our school, we use phones as part of classroom instruction. If you are not able to have students use their phone to join, perhaps computer access can be obtained through a mobile lab or a stationary lab, or the code can be dispensed to have them join on their own.

Starting Class - I have the students practice with the pickup table as they enter, selecting their journal and then putting their name on the front as we get started.

Journal Set Up - We keep two pages blank for the Table of Contents and begin gluing the Rules and Safety contract after that. I also go over a few more of the expectations while we are working, such as my journal policy (checked for a minor grade each grading period) and our bathroom procedures (two emergency passes each grading period). We go slow, working through each task together – TOC, rules, contract, calendar and vocab. We glue each, then write them on our TOC. Finally, we glue the envelope inside the back cover of the journal and label it REFERENCE. Students then fold the periodic table in half and put it inside. If this is too much for one day, you can leave off whatever you like, but we will be using all of these items for lessons later, so you will either want them in place and ready, or be prepared to take care of setting them up at a later time.

At the end of class, students are directed where to store their journals – we never take them out of the room to prevent their being lost or damaged. I bought colored plastic crates at a super store that were very inexpensive and have a different color for each class period makes them easy to recognize.

Topic: Reading Strategy & Safety Data Sheets
Day: 3
Unit: Lab Safety and Equipment

Learning Target:
I can observe and practice reading and annotating a science text, then answer questions about Safety Data Sheets.

Student Goals:
Students need to have a basic understanding of Safety Data Sheets and how to locate important information about the chemicals they will use in class and in the future.

Agenda:
A – Warmup: Annotating Reflection
L – Reading and Annotation Modeling
A – Focus Questions

Student Materials:
Copies of the lesson article for each student
Pens and highlighters for annotating
Copies of the Sentence Stems for each student
Blank paper for warm up exercise for each student

Props:
Teacher copy of the lesson article
Warmup prompt to display or written on the board
Questions and method for submitting them

Actions and Rationale:

Warmup – annotating reflection. This exercise is designed to get the students thinking about the day's activity and to give them an opportunity to share what they have previously learned. The prompt should be something like…

Tell me about a method of annotating a text you have learned. When and how do you use it?

Give the students three to five minutes to write, then you can have them discuss with partners or in groups of four. Finally, have one or two share out to the class and add whatever you like as an introduction to the day's lesson. If you want to keep points on their warm ups, have them turn them in before they leave.

Lecture – Reading and annotating a text is an important skill, especially when working with difficult subject matter. I use articles and sections from our textbook to add enrichment to our units and usually use topics that we don't cover or discuss in class.

In addition to the annotating, the students answer a few questions about the material, often questions that are thoughts and opinions and aren't something they can simply search for the answer online. You can have them answer on paper, or like me, you can have them answer in a google document through google classroom. Whichever way you choose, you will explain how they will submit their answers at the end of the lesson.

Before you get to the questions, you will model how to use your annotation strategy. There are several out there, so do a little research and choose one that is comfortable for you to use. Then, practice teaching someone how to use it before presenting it on class day.

Once you have presented how to use the annotation strategy, explain to your students how you will be assessing their performance. I do a 50-50 split, so my students get half their points from annotation on the text, and the other half from answering the questions.

Activity – time for the focus questions if there is time left in class. If not, they can be assigned to do outside of class and submitted when completed. Again, mine are submitted via google classroom, but your students can use any form (including pen and paper) that you wish.

Topic: Safety Project Building
Day: 4
Unit: Lab Safety and Equipment

Learning Target:
I can research and create a mini-poster to present to the class about one of our lab safety rules, procedures or hazards.

Student Goals:
Students will be tested over the hazard symbols, and they will need to be familiar with the location and use of safety equipment in the room, as well as the rules for being in the lab.

Agenda:
A – Warmup: Sign Up for a Topic
L/A – Safety Project: Set Up & Get Started

Student Materials:
8.5 x 11 paper for each student
Copy of the directions for each student
Markers, pens and colors

Props:
Grading Rubric for the project
Copy of the Lab Safety Rules for rule assignments

Actions and Rationale:
Warmup – assigning the topics. You can walk around and pencil the names in or pass around a list of the Lab Safety Rules and have the students put their name next to one of the rules. You can make them go down the line, or you can prearrange who is getting what rule ahead of time. You could let them choose, but it could be more difficult to ensure as many rules are covered by the students as possible. Either way, be sure you have a plan before class starts and are ready to execute as soon as attendance is taken.

Lecture / Activity – how the project works. Give the students a run-down on what you are expecting from their presentation and poster. I have included a copy of my requirements as a suggestion which can be duplicated and handed out to the students.

When I assign this project, I like to give a short introduction about risk management, which is the safety team we are all on as if we were on a job. Being unified is how we work together to prevent accidents. I also like to point out that being 99% safe may sound great, but in reality, that means that 1 to 2 students would be seriously injured in my classroom each year. That is unacceptable to me, and we will learn to be proactive this year to prevent accidents before they happen. We discuss exactly what being **proactive** and **reactive** mean and share how that might look in our class this year.

Finally, I explain how the project will work and how I will be grading them. Since this is the first project of the year, I am more interested in learning what type of students I have than whether or not they can follow directions down to the letter, so my grading is much more lenient than I let on when I announce the project.

Presentation – I tell the students they will speak two to three minutes so they will look for things to talk about. With my class sizes, they will only have about thirty seconds, and anything they don't say I will lead them into with questions or point out for myself.

I prepare a list of everything we need to have talked about, such as the shower and eyewash station, and check them off as we go to make sure we get them all. Make your own list – what is the most important to you? But, try to let the students create and share as much of the time as possible so that it doesn't turn into a lecture – if you are willing to do all the talking, they will certainly let you. It is good for them to learn to speak in front of their peers and getting their feet wet right at the beginning is a great way to introduce them to your class.

I have the presentation day set as the very next day, but you can schedule it whenever you like. I often push it down with three or four days between so that those who are absent can be given a topic to complete and we can use small bits of left-over class time to work on them if we are able. For now, give them the date they will be presenting and tell them they will need their materials ready to go by presentation day for full credit.

Prop – I tell them they will need a prop of some kind to show the class what they are talking about. For some, it will be the safety equipment, as I want to be sure we have presented all of it around the room. They can also use other equipment, or even another person, such as showing how to be properly dressed or how to use our aprons. This is a great time for pre-teaching and being proactive about our lab time when it comes, so go big here.

Mini-Poster – I tell them they must have their Safety Symbol or Hazard represented if they have one, so I specifically tell those that do that it needs to be included and I make sure it is shown to the class on presentation day. They should also include whether they are proactive or reactive, and they need to have a picture of some kind. This is a great way to find out who your project or craft-oriented learners are. On years that I have lots of them, I plan more activities they will enjoy, and when I have less, I plan less, as those who are not 'crafty' seem to ditch them, regardless if it is a grade.

Safety Presentations

Today, you will be preparing a presentation for a lab safety rule, hazard, or procedure. Being safe in Chemistry is a priority, and that means everyone must be familiar with our safety rules and procedures as well as following them as expected. Be sure to prepare a good presentation – everyone's safety this year depends on it!

~ Your presentation must include two to three minutes worth of material.
~ You will need a prop to show the class – this can be anything inside our room.
~ You need a mini-poster that includes your rule or hazard symbol and a picture.
~ On your poster, you need to designate whether your rule / hazard is proactive or reactive and be able to explain why.

© Lavish Publishing, LLC

Lab Rules and Procedures

Students will not touch anything on a lab table until instructed to do so - if a lab is set up, stay away from it before class.

Students will conduct themselves in an appropriate manner at all times - no horseplay of any kind.

Students will stay at their lab table, work with their group, and not deviate from their assigned duties / location.

Students will ensure that the floors and aisles are unobstructed during labs and activities - backpacks, etc on their chair.

Students will read and follow all written directions on the board or lab sheet.

Students will listen to instructor directions and make notes in order to follow any verbal directions.

Students who do not understand will ask for clarification before proceeding.

Students will not engage in any unauthorized experiments - no making up your own or modifying the lab in any way.

Students will not eat or drink in my classroom, which includes items brought in OR items from the lab itself - no tasting!

Students will be handling chemicals, which should not be placed near the face or touched to the skin for any reason.

Students will not sniff chemicals - if you need to detect odor, be sure to waft it gently towards your nose.

Students will not chew gum while conducting a lab. Please dispose of gum in the trash.

Students will place broken glass in the proper receptacle - the broken glass box in the back!

Students should never use a cracked or broken glass item - inspect them before you start.

Students will learn proper technique for heating a test tube - across the center and pointed away from people.

Students will be aware of hotplates / burners - never leave them unattended and avoid burn hazards.

Students will learn the safety symbols, what they stand for, and how to respond / behave in those given situations in both proactive and reactive manners. (fire hazard, corrosive hazard, toxicity hazard, eye hazard, electrical hazard, breakage hazard, burn hazard, disposal hazard, and nuclear/radiation hazard)

Students will learn the location of the emergency equipment in our classroom and how to use it. (fire blanket, eye wash, emergency button, office phone, first aid kit, and shower)

Students will not touch any of the emergency equipment unless we have an actual emergency in progress.

Students will learn the proper names of our lab equipment, along with proper handling and usage this year - see complete list for details.

Students will consult with their physician about any medical condition (i.e. contacts, allergies, etc.) and notify the instructor of any restrictions before the day of the lab.

Students will properly wear their apron when directed to do so for the entirety of the lab, or until told they can remove it.

Students will properly wear their goggles when directed to do so for the entirety of the lab, or until told they can remove them.

Students will wear their hair up or pulled back if it is long enough to go into a pony-tail holder for announced lab days.

Students will wear closed toed shoes on all appropriate lab days - no skin or socks showing.

Students will wear long pants on all appropriate lab days - no legs showing.

Students will not wear over-sized or baggy clothing or jewelry on lab days - including dangling earrings, necklaces, bracelets, or sleeves.

Students will wash their hands with soap and water after completing a lab using the large sinks at the back of the room and drying them on a towel. The small sinks are NOT for washing anything and hand sanitizer is NOT soap and water.

Students will report any accident to the instructor immediately.

Students will not remove any chemicals or materials from the lab / classroom.

Students will clean up their lab area and dispose of waste materials as instructed. (Please be sure it is trash before you throw empty containers away, though)

Students will be sure that all sinks are kept clear of debris - nothing goes in our sinks except approved liquids.

Students will be charged for any broken equipment or furniture, or damage of any kind caused by negligence, misuse, or failure to follow directions / safety procedures. (this includes desks and lab tables)

© Lavish Publishing, LLC

Topic: Safety Project Presentations
Day: 5
Unit: Lab Safety and Equipment

Learning Target:
I can present my Lab Safety Project to the class, listen as others present theirs, and be a proactive member of our Risk Management Team.

Student Goals:
Students will be tested over the hazard symbols, and they will need to be familiar with the location and use of safety equipment in the room, as well as the rules for being in the lab.

Agenda:
A – Gather your Materials and be ready to present
L/A – Lab Safety Project Presentations

Student Materials:
Access to items they might want to use as props

Props:
Ready to cover any topics that weren't assigned to a student or students who are absent
Large copies of the safety symbols in case you need them or to make an anchor chart.

Actions and Rationale:
This is a fun day, so relax and enjoy getting to know your kids. I usually have them go up in small groups, such as 'everyone who has fire' so they are not 'alone' at the front of the room. Then we go down the line and tackle each one. Again, my classes are usually thirty plus, so we really have under a minute to spend on each one. Ask them questions, make little jokes and put them at ease, and be sure to point out the major points from your list as you go if no student brings them up.

As they finish, I take their poster so I can 'grade' them – whatever they have done is what they get, and I am actually very lenient – if they turn in a poster with any decent attempt, they will get a passing grade.

I also have to keep an eye out so that no one is working on them during presentations – that would not be fair and I make them go next if I see them and then they are done. At the end, be sure to call up anyone who "hasn't gone yet." There will always be someone who was too shy to get up when their category was called and will need a little push to get up and go.

I always insist that they go and to make it easier, I say, "No one has ever died doing this." This is a great time to show them they can't simply opt out of assignments and reinforces who is making the rules in your class, so you don't have to be mean about it, but you definitely want to be firm.

Topics – this is a short list of the things I like to make sure we have covered
Safety shower
Eye Wash
Emergency Cutoff Button
Fire Blanket
Fire Extinguisher
Sharps Disposal Box
Hand Washing Sinks
Germ-ex Usage (this is NOT washing)
Aprons – how to wear and adjust
Goggles – how to wear and adjust
Clothing – what is and is not acceptable
Hair Up days – examples of hair long enough to put up
Floors – keeping walkways clear
Horseplay – it will get you set out of the lab

Topic: Lab Equipment
Day: 6
Unit: Lab Safety and Equipment

Learning Target:
I can explore and label basic lab equipment that we will be using this year in chemistry.

Student Goals:
Students will practice naming and explore how pieces of equipment are used. They will be tested over these on the unit exam and will need to know them when we begin doing labs.

Agenda:
A – Warmup: Preview Paragraph
L – Lab Equipment Modeling
A – Lab Equipment Stations

Student Materials:
Copies of the equipment inventory for each student
A way for students to get the actual name and use of equipment

Note - we use google on their phones, but if your school or class doesn't have this option, their book or a pre-printed sheet of equipment could work (you want to work them out yourself ahead of time to be aware of any issues and provide a solid method)

Props:
Lab tables set with groups of equipment
Pre-made laminated equipment label cards
Slips of paper for the warmup paragraph
Warmup question written or projected on the board

Actions and Rationale:
Warmup – preview paragraph. Have the students look over their equipment picture sheet and name one item that they have used or have seen used in the past. They should write a brief paragraph (3 to 5 sentences) explaining what the item is and how it is used / what it does. Being able to express themselves well in writing is important, so don't skimp here. I use these little assignments as **practice grades** that will convert to extra credit at the end of the six weeks. If they are weak, they get a 1, medium, they get a 2, and if they write a good paragraph, they get a 3. This tidbit of data gives me a way to track how they are doing before the test gets here.

Lecture – equipment inventory modeling. I like to use the **wash bottle** as my demo for this – I go to that table, hold it up and say "What's this?" Someone might know, but likely not, so I ask, "How can we find out?"

On the table, there are label cards with all the names on them, so I pick up the wash bottle card and show it to them. "I think this is it. How can I make sure?" I tell them that Google is my friend, so I can do a quick search, see what they are for, and even see pictures of different styles. I explain how they are going to label it on the list, and if you want, they can make a list of how each is used on the back. This may seem time consuming, but kids are smart and will quickly divide up who is searching for which item and then share with their team (I have them work in 4-somes) – you can let them, but be sure they understand they are ALL responsible for the material, knowing and turning in their own page.

For this one, I want to be sure to point out that a wash bottle holds DISTILLED WATER ONLY. I ask them to never refill my wash bottles out of the sink, as that is city water and therefore dirty – we need clean water for labs.

You can go from there, with another demo or two, you can have them do each table for so many minutes and time / rotate them by command, or simply turn them loose. I ask them to reset the cards after they have finished at the table, but you will likely have to keep reminding them as they will leave them matched and walk away. I also tell them to take a picture of the table before and after matching to study before the test, which is something we do all year long. I have my students make a **Chemistry Folder** on their phone and that is their chemistry notebook that they can take home – so learning to take pictures of everything we do will be an important documenting and study tool for them.

Activity – equipment inventory stations. I give my students the rest of class to complete the assignment, which they will put in the turn in tray when done. I don't grade them – I check off if they completed the assignment using my practice grade scores of 1, 2 or 3 after a quick scan to see how well they tried. When I return them the next day or so, I post the answers in a designated place in my room (a check my work board) and they will grade and correct their own. The students are responsible for ensuring they have the correct material to study for the exam and to use as a reference during the year. The pages are then glued in our journal, listed on the TOC, and will be one of the items on our Journal Check at the end of the grading period. I also encourage them to take a picture to go in their Chemistry Folder on their phone to study later.

Note: I leave the tables set up for several days for those who missed the activity, and students can revisit the tables later. Some play quizzing games with each other before the test using the cards and equipment, and these items could be props if your class has not presented their safety posters yet. I also post pictures in my google classroom.

Chemistry Lab Equipment

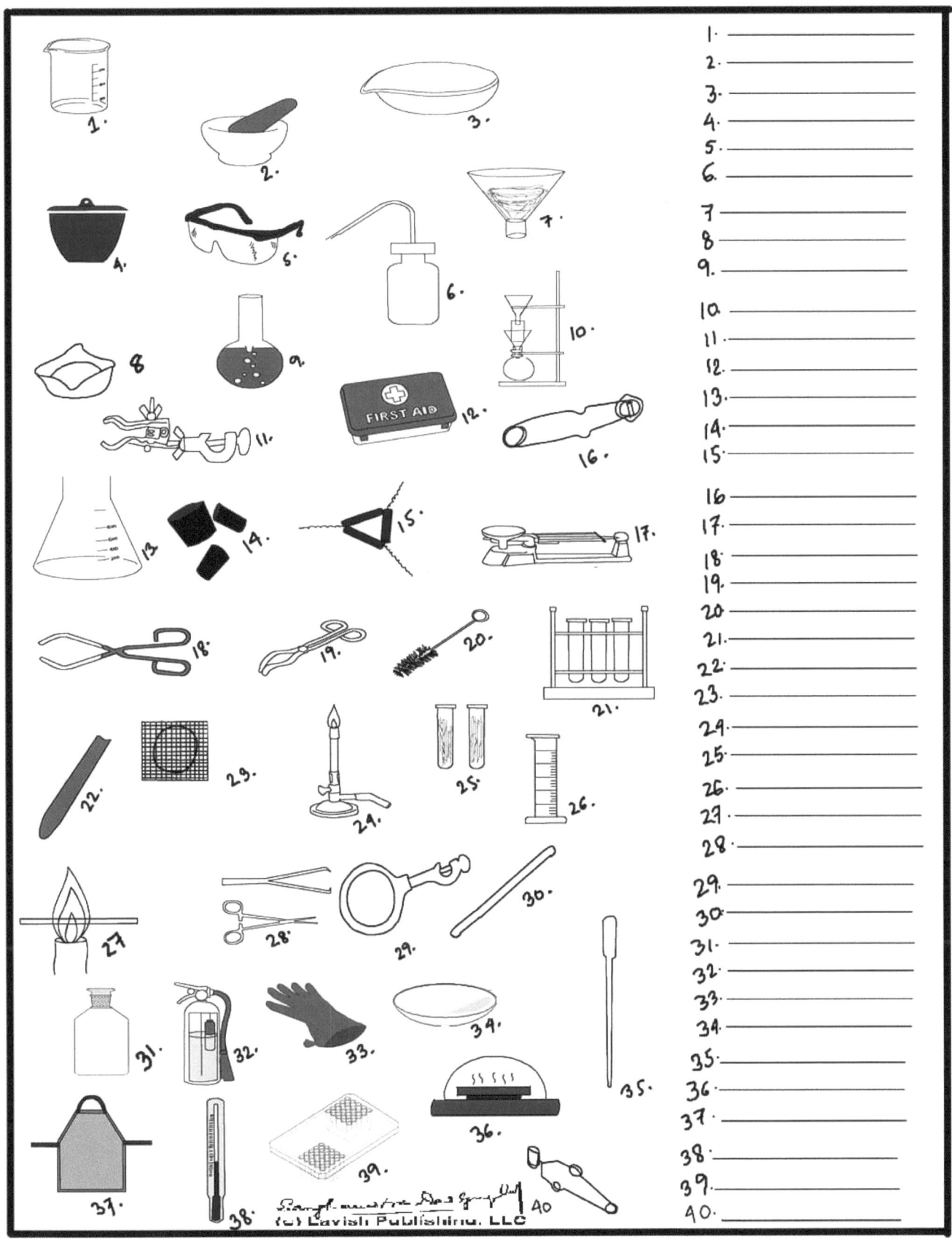

23

Equipment Stations

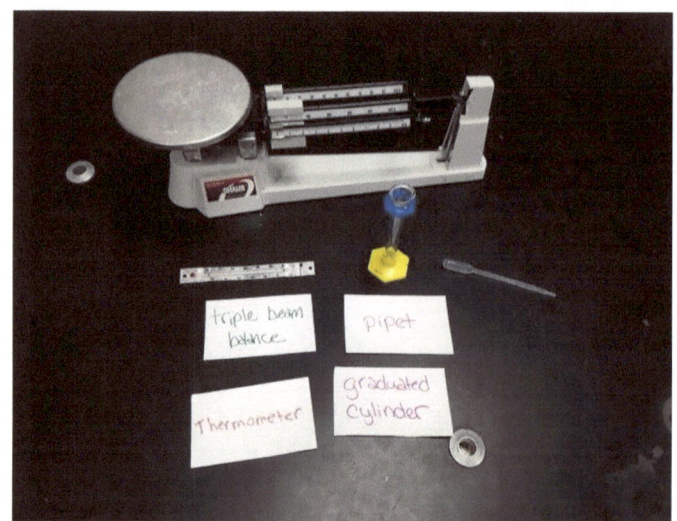

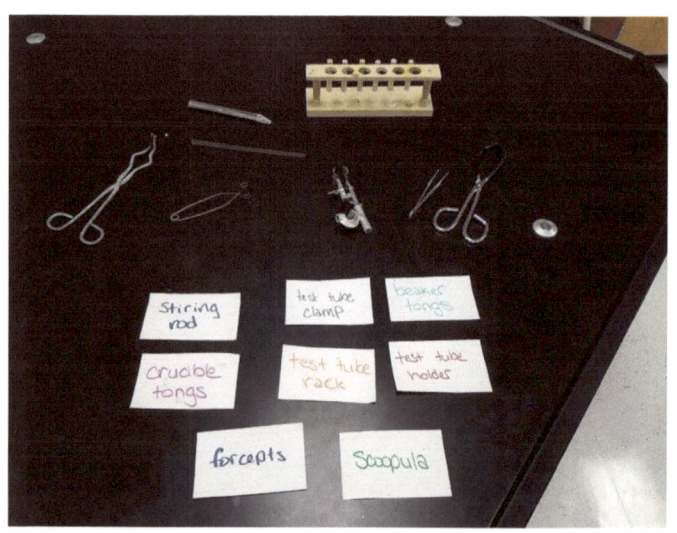
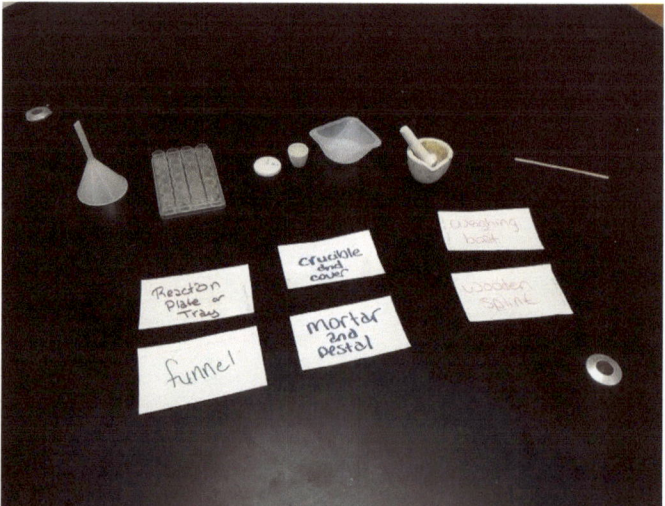
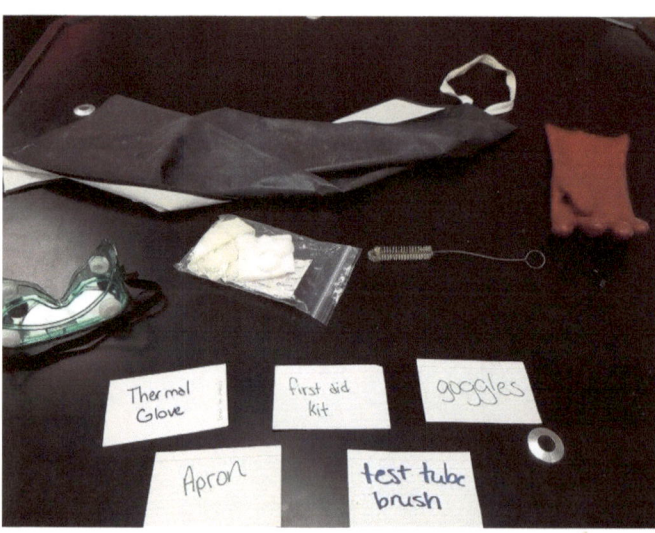
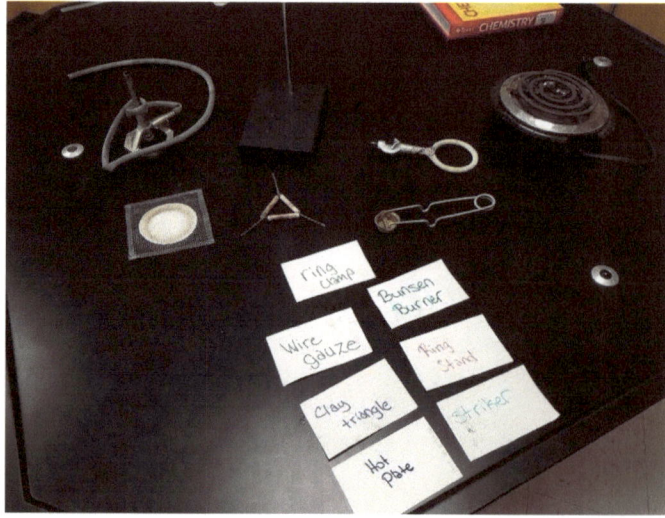

Topic: Scientific Method Lab
Day: 7
Unit: Lab Safety and Equipment

Learning Target:
I can explore and review the five main parts of a simple lab, called the helicopter lab.

Student Goals:
Students will use a simple lab to visualize, discuss and explain independent variable, dependent variable, hypothesis, and conclusion. We also discuss a control and entertain ideas of what the control in this lab might be.

Agenda:
A – Warmup: A lab in Review
L – Lab Variables and Hypothesis
A – Helicopter Lab

Student Materials:
Slips of paper for the warmup
Paper helicopters and paperclips for each
Copies of the lab report
Journals for taking notes

Props:
Anchor chart, power point, or whiteboard for lecture visuals

Actions and Rationale:
Warmup – a Lab in Review. Have the students write a half page or so defining and explaining the key terms as they understand them. This is a good time for them to practice using their vocab list if they aren't sure what the definitions are, but they should put them into their own words. You could also have them relate them to a lab that they have done in the past. I take these up and they become a practice grade – 1, 2 or 3 score depending on effort.

Lecture – lab variables and hypothesis. We spend a few minutes here reviewing – I like the dry-mix chart shown below. Once we have covered the variables, we work on the hypothesis for our lab. I go over specifically how to make it an If.. Then... statement and talk about which is the independent and dependent variables. Once they are ready, let them write their own and complete the lab.

Post Discussion – After the students have completed the lab, I go through one more recap discussion. They should all have a similar Hypothesis, and therefore they should be able to pull out the IV and DV. The control will have a larger variety of responses, and I tell them that narrowing that down with this lab is more difficult. In chemistry labs, the control will be important, as it shows us a 'non-response', as it is set aside and not experimented on.

Example board for notes and discussion:

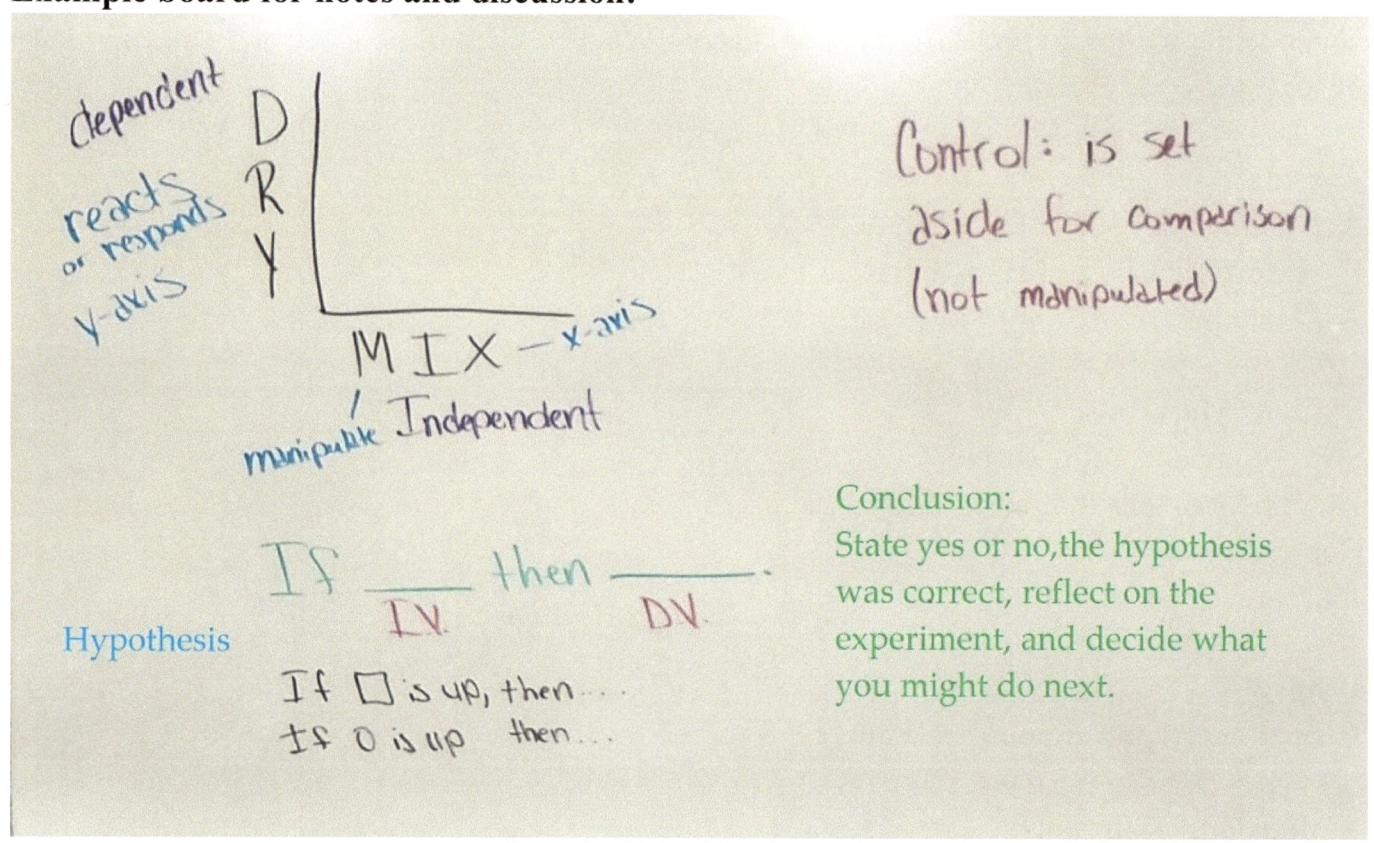

Helicopters

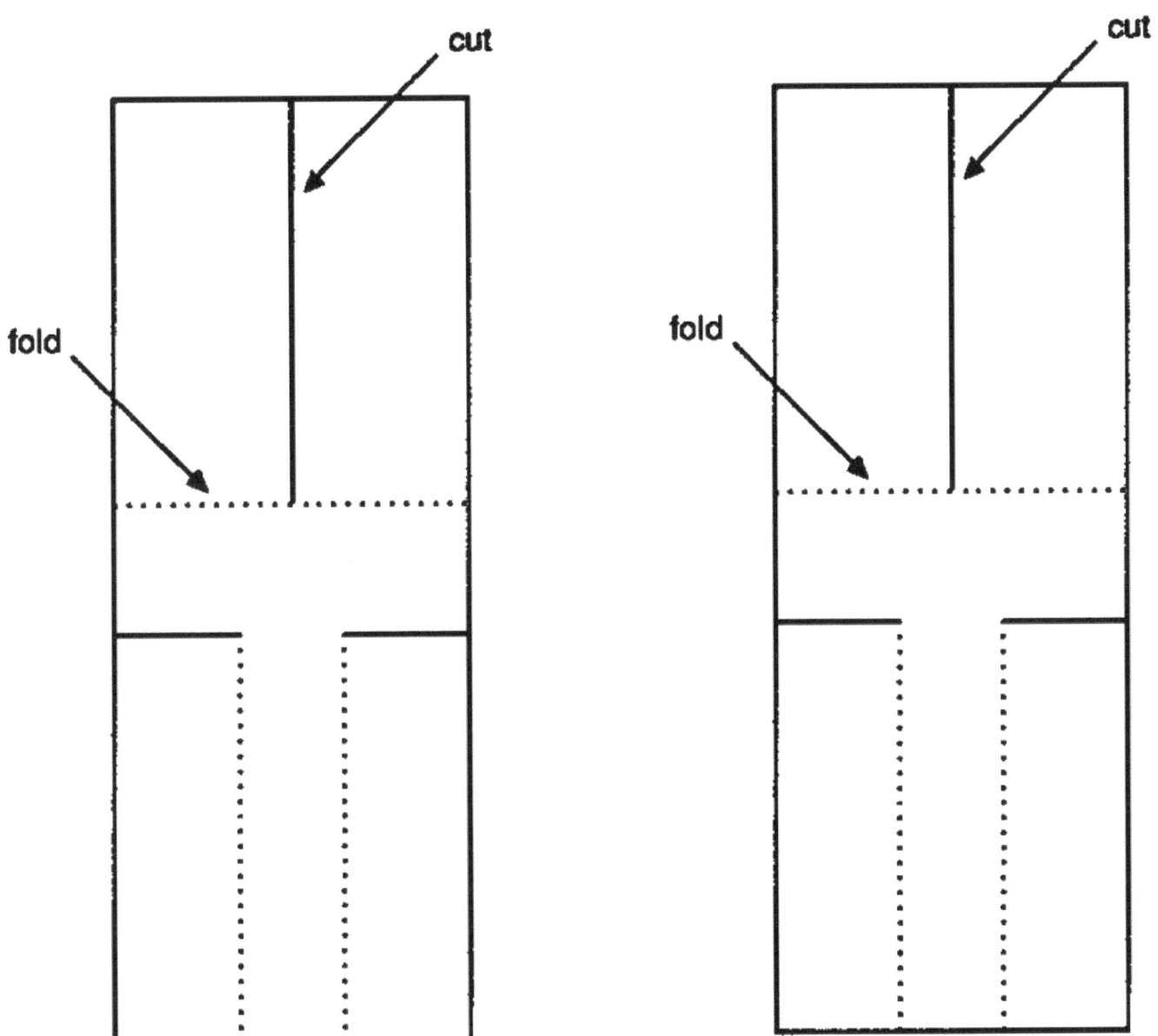

27

Helicopter Lab Report

Procedure:

1. Cut out your paper helicopter – mark one wing as "A" and the other wing as "B".
2. Write a hypothesis to fit what you think will happen when flap A is up and then you switch them so flap B is up. (Hint: think about the direction of the turning as it falls)

   If _____

   Then _____

   _____.

3. Hold the helicopter up and drop it three times with flap A up.

   Drop 1 _____ Drop 2 _____ Drop 3 _____
4. Reverse the flaps so flap B is up and repeat the process, recording your results:

   Drop 1 _____ Drop 2 _____ Drop 3 _____

Conclusions:

1. What direction did most of the flap A spins go? _____ Flap B? _____

2. Did the results prove or disprove your hypothesis? _____

   In a sentence, explain how you know: _____

   _____.

3. Name the Independent Variable _____

4. Name the Dependent Variable _____

5. What is the control in this lab? _____

   In a sentence, explain why you think so: _____

© Lavish Publishing, LLC

Topic: Video Day
Day: 8
Unit: Lab Safety and Equipment

Learning Target:
I can view videos of the consequences of labs gone wrong and write a reflection about them.

Student Goals:

Agenda:
A – Warmup Quiz
A – Video View
A – Reflection Paper

Student Materials:
Warm up Quiz slips
Paper for reflection writing

Props:
Links to videos
Power points or pictures

Actions and Rationale:
Warmup Quiz – you have three to choose from if you haven't used any of them before now. I love these little quizzes because they give me a chance to gather a little data on how our learning is going, as well as a chance to reteach the topic. The students fill out the quiz from copies or on blank slips of paper with the quiz displayed on the board. I take them up, then we go over the answers together. I grade them later for a practice grade, score 1, 2 or 3.

Videos, power point, and pictures – you can set these up as whatever you like. There are plenty of examples of how accidents happen, so have a look around and choose whether you want to go with funny, serious, or scary, as there are examples of all.

Reflection Paper – give the students about ten minutes to spend writing a full page about lab safety. A couple of good prompts might be…

Why is lab safety important to you?

Is it ok if your lab partner doesn't follow the rules in the lab? Persuade me why or why not.

Topic: Safety and Equipment Review
Day: 9
Unit: Lab Safety and Equipment

Learning Target:
I can review material for the Lab Safety and Equipment Exam.

Student Goals:
Lab Safety Rules - identify rules and examples / non-examples
Lab Equipment - be able to identify the basic equipment
Hazards - what are they, precautions, emergencies & disposal
Scientific Method - dependent, independent, hypothesis, conclusions

Agenda:
A – Warmup Quiz
A – Safety and Equipment Review
A – Lab Rule Cartoon
A – Missing work makeup
A – Safety and Equipment Vocabulary Quiz

Student Materials:
Slips for the warmup quiz
Review sheets
Progress reports or missing item slips
Copies of Vocabulary Quiz

Props:
Lab rule cartoon(s)
Sentence Stem (to display or written on the board)

Actions and Rationale:
Warmup Quiz – time to have another one and go over it at the start of class. Again, I take them up before we talk about the answers and I use them as a reteach opportunity and a practice grade, scored as 1, 2 or 3.

Review – I don't do paper reviews in class often, but I do like games and activities. If the equipment is still sitting on the tables, students can quiz each other. If there is a place to lay out the posters from the presentations, or if you want to make a selection of them to post on the board, students can also look them over and discuss to review the rules and symbols. Finally, they should also go over the Helicopter Lab and scientific method.

Lab Rules Cartoon – this is one activity that could be used on any day but would also make a good review. It also has a number of options how to proceed – it can be a find and write, a partner point and talk (and talking about the material is always good), or even a group activity. An example stem might be, "I see ____, who is breaking / obeying _____ rule because they are _____." Mix it up and see what you like, as there will be lots of these types of activities through the year to choose from.

Makeup Day – this is a great day for students to work on and hand in anything they are missing, so give them a list of those items as they come in or while they are taking their quiz. Then they can choose to work on the review or their missing work, both of which are beneficial.

Lab Safety and Equipment Vocab Quiz – this is also a good day to give the vocab quiz, which will only take a few minutes. I have them put all their stuff under their desk except a writing item to take the quiz, and I tell them to be sure not to copy off their neighbor since they have a different quiz. You could also give this to them on a different day or along with the exam tomorrow.

Chemistry Lab Rules Cartoon

Topic: Safety and Equipment Exam
Day: 10
Unit: Lab Safety and Equipment

Learning Target:
I can demonstrate my understanding of lab safety and equipment on a written exam.

Student Goals:
Lab Safety Rules - identify rules and examples / non-examples
Lab Equipment - be able to identify the basic equipment
Hazards - what are they, precautions, emergencies & disposal
Scientific Method - dependent, independent, hypothesis, conclusions

Agenda:
A – Red button day
A – Lab Safety and Equipment Exam

Student Materials:
Scantron or answer document
Exam copies

Props:

Actions and Rationale:
Red Button Day – classrooms run on routines, and this is one of my most hard fast, so this short and quick exam is a great time to introduce my students to my testing procedures. Basically, I have a red, yellow and green button up on my board next to the date, which is almost always on yellow. Yellow means that the students can be on their phones for APPROVED activities.

On red button day, ALL of their personal stuff goes in a designated location. I have a couple of folding tables at the front of my room that are set to about a foot tall, so they are short. Underneath, I have rugs because many kids don't like putting their bags on the floor underneath. Their bags, books, purses, etc. go on or under those tables and we do not move further until everyone has complied.

Their phones go either in their bag or on a charger, which I have two stations in my room – one that is a power strip where they use their charger, and a second set up with my personal chargers that I bought for the class to use – iPhone and android.

After we are seated and ready, I give them my testing rules – starting with the phones. We do the "pocket pat" so they can check to make sure their phones are put away and not accidently back in their pocket, and I tell them what is going to happen if they don't. I like to have them stay seated and working until everyone is finished (I do allow them to slip out to the bathroom QUIETLY). I pick up all the exams at the end, and they are not allowed to get ANY of their stuff until I have all the exams in hand, and I dismiss them to get their stuff.

Phone penalty – if I see them with a phone – using it or not – until they have been dismissed, they get a 1 on the exam (a signal that they have broken testing rules). On the first offense, I let them take the exam again during tutorials the following week and contact their parents to let them know what happened and make the offer of a retake. After that, they get a 1 and it stands. This isn't a game, and they need to learn the consequences of breaking test security.

Yes, I am very strict on this. I am not naive, and I know that kids use their phones to cheat in an unlimited number of ways when given the chance. For this one small piece of my class, I want them to show me what they know on their own, and I don't want them sharing copies of my test in any way. If you are strict and diligent, the number of problems will be far less in the long run. Be clear and up front with your expectations and stick to your rules. The first few exams they may whine and complain, but it should die down. Anyone who is over the top or causes disruption of the test over it gets a phone call home so I can chat with their parents about why their student can't follow the rules like everyone else.

For the **answer documents**, we use scantrons, which makes grading super easy, but if you don't have access, you can always have them fill out their answers on a strip of paper, which will make them easier to grade. I also like to provide each student a copy of the exam, which I always print and shrink to fit on a single page. I like to do this so they can practice writing on the test and using test taking strategies as they work. I also keep them locked up before the test, and after they are used for the rest of the year in case we need to go back and look at someone's test, so I do ask them to put their names on them.

How ever you go, make sure that you get ALL of your tests and answer documents back before students get their stuff or leave the room.

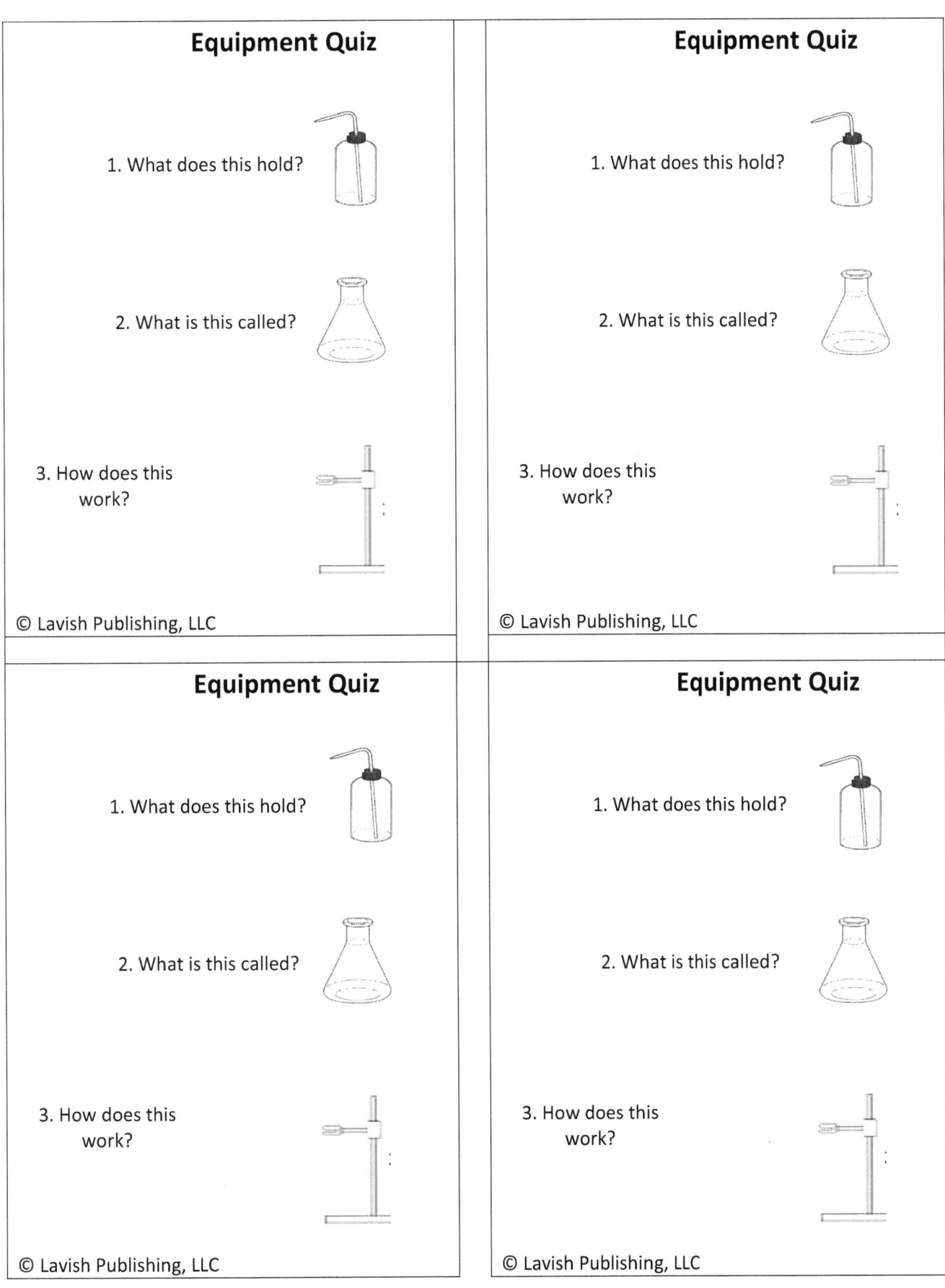

Hazard Quiz

1. This warns against…

2. These protect against…

3. Where do we dispose of this?

© Lavish Publishing, LLC

Hazard Quiz

1. This warns against…

2. These protect against…

3. Where do we dispose of this?

© Lavish Publishing, LLC

Hazard Quiz

1. This warns against…

2. These protect against…

3. Where do we dispose of this?

© Lavish Publishing, LLC

Hazard Quiz

1. This warns against…

2. These protect against…

3. Where do we dispose of this?

© Lavish Publishing, LLC

Scientific Method Quiz

1. What are you testing by manipulating it?

2. What needs to be written as a testable "if/then" statement?

3. What is set aside and never tested on?

© Lavish Publishing, LLC

Scientific Method Quiz

1. What are you testing by manipulating it?

2. What needs to be written as a testable "if/then" statement?

3. What is set aside and never tested on?

© Lavish Publishing, LLC

Scientific Method Quiz

1. What are you testing by manipulating it?

2. What needs to be written as a testable "if/then" statement?

3. What is set aside and never tested on?

© Lavish Publishing, LLC

Scientific Method Quiz

1. What are you testing by manipulating it?

2. What needs to be written as a testable "if/then" statement?

3. What is set aside and never tested on?

© Lavish Publishing, LLC

Form A Lab Safety & Equipment Vocabulary Quiz
© Lavish Publishing, LLC

1		___ are worn to protect our eyes when we are using chemicals, glassware, or fire (all labs in chemistry probably require them).
2		___ is an educated guess.
3		___ is done to obtain more accurate data.
4		___ is evidence that is found by a direct observation, experience, or situation, which makes it more believable.
5		___ is to use our resources sparingly or in a non-wasteful manner. It also refers to some of our laws in science that state that certain things cannot be created or destroyed, such as matter or energy.
	word choices	(A) conservation (B) empirical evidence (C) hypothesis (D) repeated trials (E) safety goggles

Form B Lab Safety & Equipment Vocabulary Quiz
© Lavish Publishing, LLC

1		___ means to throw something away or get rid of it - we always must do this in the proper manner to prevent contaminating our water supply and landfills.
2		___ would be looking at all of the facts and lining up those facts in such a way that you come to a documented conclusion.
3		___ is the ability to burn or catch fire.
4		A ___ is a flask with a narrow tube from the mouth down to a fat round base.
5		A ___ is a laboratory instrument used to transport a measured volume of liquid, similar to an eye-dropper.
	word choices	(A) disposal (B) flammability (C) Florence Flask (D) logical reasoning (E) pipette

Form C **Lab Safety & Equipment Vocabulary Quiz**
© Lavish Publishing, LLC

1		A ___ is a piece of laboratory glassware or plastic ware used to accurately measure out volumes of chemicals or other liquids. They are generally more accurate and precise for this purpose than flasks.
2		A ___ is anything that can be changed or changes.
3		A ___ is used to hold chemicals in storage - one should NEVER pour things into or stick items inside of them to prevent contamination.
4		A ____ has a short neck from the mouth to a body that flares out from a narrow top to a wide base.
5		A ____ is unexpected and causes property damage or personal injury.
	word choices	(A) accident (B) Erlenmeyer Flask (C) graduated cylinder (D) reagent bottle (E) variable

Form D **Lab Safety & Equipment Vocabulary Quiz**
© Lavish Publishing, LLC

1		A ____ is used to record data and observations for lab activities.
2		A piece of lab safety equipment that can dump water on you entire body in case of an emergency is the ____ located in the back of our classroom.
3		An ____ is something that we notice or make note of while performing a task or experiment.
4		An _____ value is found in a lab.
5		Chemicals with high ____ can cause skin irritation and burns, as well as making stains on and holes in clothing and other items.
	word choices	(A) corrosiveness (B) experimental (C) lab journal (D) observation (E) safety shower

Lab Safety and Equipment Exam

Match the following Parts of the Scientific Method in questions 1 – 5.
A. Independent Variable   C. Hypothesis   E. Control
B. Dependent Variable   D. Conclusion

1. Which is the one that sums everything up and explains what you found?

2. Which is the one you were measuring as it reacted?

3. Which is the one that was set aside and never tested on?

4. Which is the one you were testing by manipulating it?

5. Which is the one that must be a testable *"if / then"* statement?

Match the following Safety Symbols in questions 6 – 10.
A.    B.    C.

D.    E.

6. Which symbol means the substance is corrosive?

7. Which symbol would go with the sharps disposal box?

8. Which symbol would prevent the need of the eye-wash stations?

9. Which symbol warns you not to ingest the chemical?

10. Which symbol warns you about flammability?

© Lavish Publishing, LLC

Match the following pieces of equipment in questions 11 – 15.

A. 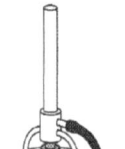 B.  C.  D.  E.

11. Which should not be left unattended when in use?

12. Which is the beaker?

13. Which is for DISTILLED water only?

14. Which is called a graduated cylinder?

15. Which is the Erlenmeyer Flask?

Match the following pieces of equipment in questions 16 – 20.

A.  B. C. 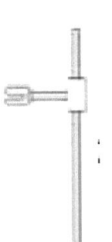 D. E.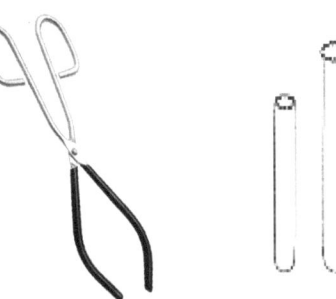

16. Which is tongs?

17. Which one measures temperature?

18. Which is the pipette?

19. Where is the test tube?

20. Which is made from a metal pole that things clamp to?

© Lavish Publishing, LLC

Lab Equipment Answers

1. beaker
2. mortar and pestle
3. evaporating dish
4. crucible and cover
5. safety goggles
6. wash bottle
7. funnel
8. weigh boat
9. Florence flask
10. ring stand
11. test tube clamp
12. first aid kit
13. Erlenmeyer flask
14. rubber stoppers
15. clay triangle
16. striker
17. triple beam balance
18. beaker tongs
19. crucible tongs
20. test tube brush
21. Test tube rack
22. scoopula
23. Wire mesh
24. Bunsen burner
25. Test tubes
26. Graduated cylinder
27. Wooden splint
28. forceps
29. Ring clamp
30. Glass stirring rod
31. Reagent bottle
32. Fire extinguisher
33. Thermal glove
34. Watch glass
35. pipette
36. hotplate
37. apron
38. thermometer
39. Reactivity tray
40. Test tube holder

Quizzes Answer Keys

Equipment Quiz

1. What does this hold?
distilled water

2. What is this called?
Erlenmeyer flask

3. How does this work?
A ring stand has items clamped to it for fire use

© Lavish Publishing, LLC

Hazard Quiz

1. This warns against...
fire (flammability hazard)

2. These protect against...
getting chemicals and glass in your eyes (eye hazard)

3. Where do we dispose of this?
broken sharps container

© Lavish Publishing, LLC

Scientific Method Quiz

1. What are you testing by manipulating it?
independent variable

2. What needs to be written as a testable "if/then" statement?
hypothesis

3. What is set aside and never tested on?
the control

© Lavish Publishing, LLC

Lab Safety and Equipment Vocabulary Quiz Answers

Form A		Form B		Form C		Form D	
1	E — safety goggles	1	A — disposal	1	C — graduated cylinder	1	C — lab journal
2	C — hypothesis	2	D — logical reasoning	2	E — variable	2	E — safety shower
3	D — repeated trials	3	B — flammability	3	D — reagent bottle	3	D — observation
4B	B — empirical evidence	4	C — Florence flask	4	B — Erlenmeyer flask	4	B — experimental
5	A — conservation	5	E — pipette	5	A — accident	5	A — corrosiveness

Lab Safety and Equipment Exam Answers

1. D	6. E	11. A	16. D
2. B	7. C	12. E	17. A
3. E	8. B	13. B	18. B
4. A	9. D	14. D	19. E
5. C	10. A	15. C	20. C

About the Author

Born and raised in West Texas, Sammie Jacobs aspired to be a teacher at an early age but did not achieve her dream until the age of 38 when she earned her composite science certification in 2008.

Answering the call of the classroom, she went to work at a local high school teaching Chemistry. Through all the years, she loved the subject and her students. Combining her dedication to both, Sam continuously searched for ways to improve her instruction and meet her students varied needs, leading her to create and construct many of her materials.

Mentoring new teachers as they came into the district and her department, Sam could see the importance of helping those new to her beloved field. First, she began to build the files that would become the foundation of this series. Later, she realized that putting those lesson plans and tools into the hands of others could mean a great deal to her colleagues and countless students, eventually even those across the country or around the world.

Always striving for more and looking for the best in herself and those around her, Sammie Jacobs is releasing this complete version of her Chemistry lessons – 17 units in all, which will be available for purchase at a nominal fee in paperback format. These files contain everything needed to plan and execute a solid foundational year of Chemistry for any teacher, be it veteran or novice.

But Sam also knows there is more that can be done, so she is also founding a teacher community group on Facebook. Open to everyone who works as a science educator, this is Sam's legacy, her dream coming true. If you are a veteran with advice to share or a novice looking for a helpful hand, come and join and let us grow together. In the end, Sam hopes many will find these tools useful and many more will be inspired to reach for their dreams, whatever they might be…

Units in the Teaching Chemistry in a Diversified Classroom Series

Lab Safety & Equipment
Properties of Matter
Periodic Table
Atomic Structure
Electrons in Atoms
Ionic Bonding
Covalent Bonding
Names & Formulas
Equations & Reactions
Calculations
Stoichiometry
States of Matter
Gases
Solutions
Acids & Bases
Thermochemistry
Nuclear Chemistry

Follow the Teaching Chemistry Series on Facebook –
https://www.facebook.com/SammieJacobsChemistry/?

Join the Chemistry Professional Learning Community on Facebook –
https://www.facebook.com/groups/TeachingChemistryPLC/